British Bird Puzzle Quiz Book

British Bird Puzzle Quiz Book

Sean Nicholson

Mombolo Publishing
2017

First Printing: 2017

ISBN 978-0-244-63690-6

Mombolo Publishing
Oxford,United Kingdom

www.MomboloPublishing.com

Contents

Preface

This book is my attempt to have some fun and hopefully entertain others. In those dark evenings when bird watchers can only listen out for owls calling, this is a diversion to amuse. I have tried to find many different ways to set puzzles and quiz questions that maintain the reader's interest. There is no prize for completing them all, just the satisfaction, or frustration of having tried all the different questions.

Questions are on the right hand pages, and the answers are on the next page i.e. turn the page to find the answers, but only after you have answered all the questions.

I hope you enjoy it.

Sean Nicholson
September 2017

Warm up General Knowledge

1. A female Ring Ouzel could be confused with which of its close relatives?

2. What is a group of Owls called?

3. How many species of Swan occur naturally in the UK?

4. Which bird seen in the UK is unique in having an asymmetrical plumage pattern?

5. Which bird is the symbol of the Royal Society for the Protection of Birds (RSPB)?

6. What is a Green Plover more commonly known as?

7. What is the smallest British bird?

8. Who founded the Wildfowl and Wetlands Trust (WWT)?

9. Which is more common in the UK, the Rook or the Magpie?

10. Which bird has the Latin name Troglodytes troglodytes?

Warm Up General Knowledge – Answers

1. A female Blackbird.

2. A Parliament of Owls.

3. Three – Mute, Bewick's, Whooper.

4. The Wryneck, on the back of its neck.

5. The Avocet.

6. A Lapwing.

7. The Goldcrest.

8. Peter Scott.

9. Rook – nearly 1 million pairs, versus half a million for Magpie.

10. The Wren.

Word Search

In the grid below are the start of eight bird names matching the second half of their name in the list after the grid e.g. if it said Pipit then you might search for Tree or Rock in the grid below.

X	V	T	E	H	T	B	G	E	O
L	L	I	T	T	L	E	O	C	P
S	O	P	R	A	N	E	L	O	K
H	P	A	N	I	O	W	Y	A	C
D	H	O	T	W	N	M	S	L	Q
T	O	C	T	O	Z	G	M	S	S
W	O	E	M	T	B	R	E	N	T
A	D	Z	E	W	E	J	A	D	R
H	E	L	O	L	A	D	A	U	E
S	D	F	G	R	E	E	N	V	E

1. _____ Plover
2. _____ Tit
3. _____ Goose
4. _____ Sparrow
5. _____ Crow
6. _____ Gull
7. _____ Woodpecker
8. _____ Flycatcher

Word Search - Answers

	L	I	T	T	L	E		C
S			R					O
	P			I				A
	H	O			N			L
	O		T			G		
	O			T	B	R	E	N	T
	D				E			D	R
	E					D			E
	D		G	R	E	E	N		E

1. Ringed Plover
2. Coal Tit
3. Brent Goose
4. Tree Sparrow
5. Hooded Crow
6. Little Gull
7. Green Woodpecker
8. Spotted Flycatcher

Bird Dates

Match the event to the correct year.

1. Collared Doves started to breed in the UK.

2. Little Egrets started to breed in the UK.

3. Ospreys started breeding at Loch Garten.

4. Corncrakes start to be reintroduced in East Anglia.

5. Slimbridge Wildfowl Centre opened to the public.

6. Red Kite introduction program started in England and Scotland.

7. White Tailed Eagle reintroduction program starts on the island of Rhum (now Rùm).

8. Minsmere becomes an RSPB reserve.

1946, 1947, 1956, 1959, 1975, 1989, 1996, 2003

Bird Dates – Answers

1. Collared Doves started to breed in the UK. 1956

2. Little Egrets started to breed in the UK. 1996

3. Ospreys started breeding at Loch Garten. 1959

4. Corncrakes start to be reintroduced in East Anglia. 2003

5. Slimbridge Wildfowl Centre opened to the public. 1946

6. Red Kite introduction programme started in England and Scotland. 1989

7. White Tailed Eagle reintroduction program starts on the island of Rhum (now Rùm). 1975

8. Minsmere becomes an RSPB reserve. 1947

Breeding Seabird Population Size

Place the following in order of breeding population size from the most rare to the commonest breeding seabird.

1. Arctic Skua

2. Common Tern

3. Roseate Tern

4. Puffin

5. Razorbill

6. Guillemot

7. Gannet

8. Common Gull

9. Kittiwake

10. Great Black-backed Gull

11. Mediterranean Gull

Breeding Seabird Population Size - Answers

1. Roseate Tern – 80 pairs

2. Mediterranean Gull – 600 pairs

3. Arctic Skua - 2000 pairs

4. Common Tern – 10,000 pairs

5. Great Black-backed Gull – 17,000 pairs

6. Common Gull – 50,000 pairs

7. Razorbill – 110,000 pairs

8. Gannet – 220,000 pairs

9. Kittiwake – 370,000 pairs

10. Puffin – 580,000 pairs

11. Guillemot – 880,000 pairs

Parts of a Bird

Match the letter to the plumage name from the list below

1. Upper tail covert

2. Lesser coverts

3. Middle coverts

4. Greater coverts

5. Secondaries

6. Primaries

7. Ear covert

8. Under tail coverts

Parts of a Bird - Answers

1. Upper tail covert = F

2. Lesser coverts = B

3. Middle coverts = C

4. Greater coverts = D

5. Secondaries = E

6. Primaries = G

7. Ear covert = A

8. Under tail coverts = H

Even Chances

1. Which duck has a pale blue forewing – the Male Pintail or Male Shoveler?
2. Which Swan has a triangular shaped wedge of yellow on its beak – Bewick's or Whooper?
3. A Diver with a large white patch on its side is a –Red-throated Diver or Black-throated Diver?
4. Which has the longer primary wings – Chiffchaff or Willow Warbler?
5. Which is an introduced duck species – Gadwall or Mandarin?
6. A Brent Goose in the south-east of England is probably a – Dark-bellied or Light-bellied?
7. A black and white woodpecker in Scotland is probably a –Lesser Spotted or Great Spotted?
8. A black bird with a white forehead, sitting on a lake is probably which type of bird – a Coot, or a Moorhen?
9. Do the male and female look similar in Robins or Yellowhammers?
10. A Pipit seen in winter is probably a – Meadow Pipit or a Tree Pipit?

Even Chances – Answers

1. Male Shoveler

2. Whooper

3. Black-throated Diver

4. Willow Warbler

5. Mandarin

6. Dark-bellied

7. Great Spotted

8. A Coot

9. Robins

10. Meadow Pipit

Not in Ireland

Which of these birds are unlikely to be seen in Ireland? There are multiple correct answers.

1. Hawfinch

2. Water Rail

3. Yellow Wagtail

4. Barn Owl

5. Tawny Owl

6. Long-eared Owl

7. Water Pipit

8. Tree Pipit

9. Corn Bunting

10. Yellowhammer

11. Pied Flycatcher

12. Siskin

13. Red Grouse

14. Redstart

15. Nightingale

Not in Ireland - Answers

These birds do not normally occur in Ireland.

1. Hawfinch

2. -

3. Yellow Wagtail

4. -

5. Tawny Owl

6. -

7. Water Pipit

8. Tree Pipit

9. Corn Bunting

10.-

11.Pied Flycatcher

12.-

13.-

14.Redstart

15.Nightingale

More Common

Which has the largest British breeding population?

1. Redpoll or Siskin?

2. Greenfinch or Goldfinch?

3. Ringed Plover or Little Ringed Plover?

4. Golden Plover or Shelduck?

5. Pochard or Eider?

6. Mute Swan or Tawny Owl?

7. Long-eared Owl or Short-eared Owl?

8. Water Pipit or Rock Pipit?

9. Tree Pipit or Grey Wagtail?

10. Cuckoo or Water Rail?

11. Bearded Tit or Crested Tit?

12. Mistle Thrush or Song Thrush?

<u>More Common – Answers</u>

1. Siskin

2. Greenfinch

3. Ringed Plover

4. Golden Plover

5. Eider

6. Tawny Owl

7. Long-eared Owl

8. Rock Pipit

9. Tree Pipit

10. Cuckoo

11. Crested Tit

12. Song Thrush

Latin Names

What is the English name for each bird's Latin name below?

1. Pica Pica

2. Turdus merula

3. Puffinus puffinus

4. Erithacus rubecula

5. Milvus milvus

6. Tetrao tetrix

7. Troglodytes troglodytes

8. Falco Subbuteo

9. Cinclus cinclus

10. Gavia immer

11. Passer montanus

12. Anas crecca

Latin Names – Answers

1. Magpie

2. Blackbird

3. Manx Shearwater

4. Robin

5. Red Kite

6. Black Grouse

7. Wren

8. Hobby

9. Dipper

10. Great northern Diver

11. Tree Sparrow

12. Teal

Named After

Name all the British birds that have someone's name as part of their own name e.g. Cetti for Cetti's Warbler.

The answer page has 41 names.

Named After – Answers

Allen	Macqueen
Audouin	Marmora
Baillon	Montagu
Baird	Moussier
Barrow	Naumann
Bewick	Pallas
Blyth	Person
Bonaparte	Radde
Bonelli	Ross
Brünnich	Rüppell
Cabot	Sabine
Cetti	Savi
Cory	Steller
Cretzschmar	Swainson
Eleonora	Swinhoe
Fea	Sykes
Forster	Temminck
Franklin	Tengmalm
Hume	White
Lady Amherst	Wilson
Leach	

British Bird List

1. Which organisation maintains the British List?

2. What is the first bird on the British List?

3. What is the last species on the British List?

4. To the nearest hundred, how many species are on the British List?

5. The 'official' British List only contains species from which categories? Possible categories are A to F.

6. What does category F represent?

7. Which of these is not on the official British List: the Sandhill Crane, Yellow-nosed Albatross, Hawk Owl, or Tawny Eagle?

8. Which parts of the UK are not covered by the British List?

British Bird List – Answers

1. The British Ornithologists' Union

2. Mute Swan

3. Wilson's Warbler

4. 600

5. A,B,C

6. Bird species recorded before 1800

7. Tawny Eagle

8. Northern Ireland, Isle of Man and the Channel Islands

Technical Term

Give the technical name for:

1. Two birds preening each other.
2. Bare skin by the nostrils seen in hawks and pigeons.
3. Birds soaring in a spiral motion on a thermal
4. Born helpless, normally naked and blind.
5. Name for the flight feathers made up of primary and secondary feathers together.
6. A protein that feathers are made of.
7. A Female having multiple male mates.
8. A sharp knob on a chick's beak that helps it to hatch by breaking open the egg.
9. A group of eggs laid in a nest.
10. Active at dusk and dawn.

Technical Term – Answers

1. Allopreening
2. Cere
3. Kettle
4. Altricial
5. Remiges
6. Keratin
7. Polyandry
8. Egg tooth
9. Clutch
10. Crepuscular

Red Birds

List as many birds as you can whose name starts with "red" on the British list. Some are common, some rare.

The answer page has 25 names on its list!

Red Birds – Answers

Red-breasted Goose
Red-crested Pochard
Redhead
Red-breasted Merganser
Red Grouse
Red-legged Partridge
Red-throated Diver
Red-billed Tropicbird
Red-necked Grebe
Red Kite
Red-footed Falcon
Red-necked Stint
Redshank
Red-necked Phalarope
Red-necked Nightjar
Red-eyed Vireo
Red-backed Shrike
Red-rumped Swallow
Red-breasted Nuthatch
Red-throated Thrush
Redwing
Red-flanked Bluetail
Red-breasted Flycatcher
Redstart
Red-throated Pipit

Do You Know?

1. Which British bird has the highest rate of adultery?

2. Which British Bunting never travels, resulting in individuals only 30km apart sometimes singing in a different dialect?

3. Which British bird has the largest difference in size between male and female?

4. Which bird is the wild ancestor of the feral pigeon?

5. Which British bird lives the longest?

6. Which male bird can be either a faeder, a satellite, or an independent?

7. Which irregular winter visitor is often found on berry producing bushes in car parks and other urban locations?

8. The seed crop of which tree prompts migrating Brambling to travel in search of it?

Do You Know? – Answers

1. Reed Bunting

2. Corn Bunting

3. Sparrowhawk

4. Rock Dove

5. Manx Shearwater

6. Ruff – linked to breeding strategy

7. Waxwing

8. Beech tree

Other Names

Below are some formerly used and local names for birds.
What is the common name used for each one?

1. Fen Sparrow

2. Bald Buzzard

3. Waterhen

4. Horse Lark

5. Bee Hawk

6. Sea Pie

7. Cairn Bird

8. Little Peter

9. Nine Killer

10. Water Swallow

Other Names – Answers

1. Reed Bunting

2. Osprey

3. Moorhen

4. Corn Bunting

5. Honey Buzzard

6. Oystercatcher

7. Ptarmigan

8. Storm Petrel

9. Red-backed Shrike

10. Sand Martin

Unique Birds

1. Which is the only migratory Pigeon or Dove in the UK?
2. Which British bird lays red eggs?
3. Which bird is endemic to Britain?
4. Which British bird has 10 tail feathers?
5. This warbler migrates south the long way – round the east, not the west, of the Mediterranean. Which is it
6. Which bird moults all its feathers twice a year?
7. Which is the only globally extinct bird to have once bred in the UK?
8. Which British wader has no hind toe?

Unique Birds – Answers

1. Turtle Dove
2. Cetti's Warbler
3. Scottish Crossbill
4. Cetti's Warbler
5. Lesser Whitethroat
6. Willow Warbler
7. Great Auk
8. Sanderling

Migration

1. After breeding I head north up the Bay of Biscay touching the UK, then after moulting I go south to the coast of West Africa before returning to the Mediterranean to breed again. What am I?

2. Which Dunlin subspecies breeds in Britain but migrates south in winter?

3. Which game bird from the family Phasianidae migrates to and from Britain each year?

4. Which duck is a summer migrant?

5. Where do British Ospreys spend the winter?

6. Which British bird migrates to the Antarctic Ocean each year?

7. Which country do British Swallows spend their winter visiting?

8. Which two breeding warblers do not migrate each year?

Migration – Answers

1. Balearic Shearwater

2. The subspecies schinzii

3. Quail

4. Garganey

5. West Africa

6. Artic Tern

7. South Africa

8. Dartford & Cetti's.

Where in Britain?

1. Which region of Britain does the subspecies of Wren known as troglodytes occur in?

2. Which two counties make up the stronghold for the Cirl Bunting?

3. Which region of the UK has breeding Goldeneye?

4. Where might you see a Short-toed Treecreeper?

5. Which county in England do Choughs breed in?

6. Which county in England is the stronghold for wintering Bean Geese?

7. Where in Wales would you go to find breeding Roseate Terns?

8. Where is the large ornithological gathering called Birdfair held every year?

9. Which region of the UK is the stronghold for Egyptian Geese?

10. Where are the Insh Marshes?

Where in Britain? – Answers

1. South-East England

2. Devon and Cornwall

3. Central Scotland

4. Channel Islands

5. Cornwall

6. Norfolk

7. Anglesey

8. Rutland Water

9. East Anglia

10. Scotland, near Aviemore

True or False?

1. Garden Survey data shows there is a correlation between eye size and time of first arrival to a garden each morning.
2. Yellow Wagtails now spend a week longer in the UK than they did in the 1960s.
3. Sand Martins do not form pairs, but breed in small groups.
4. The Mistle Thrush aggressively attacks intruders near its nest, so smaller birds nest nearby to get protection from it.
5. Dippers feed by walking underwater along the riverbed.
6. Young Reed Warblers have spots on their tongues.
7. Birds can see ultraviolet light.
8. Fulmars fly all their life, never landing except to breed.

True or False – Answers

1. True – Blackbirds arrive first.

2. True – They arrive earlier and leave later.

3. False – Dunnocks do this.

4. False – Fieldfares do this

5. True

6. True

7. True

8. False – Swifts do this

Group Names

Groups of birds are known by special names, depending on their type. Complete the following

1. A deceit of

2. A tok of

3. A murder of

4. An exaltation of

5. A bellowing of

6. A fling of

7. A watch of

8. A spring of

9. A descent of

10. A herd of

11. A train of

<u>Group Names – Answers</u>

1. Lapwings

2. Capercailles

3. Crows

4. Larks

5. Bullfinches

6. Dunlins

7. Nightingales

8. Teal

9. Woodpeckers

10. Wrens

11. Jackdaws

Bird Populations

1. Why did Whitethroats decline by 90% in 1968?

2. Which is larger population, wintering Black-necked Grebe or wintering Slavonian Grebe?

3. Roughly how many pairs of Barn Owls breed in Britain – 4000, 40,000, or 400,000?

4. Which has the smallest population - Blue Tit, Coal Tit, Great Tit, Marsh Tit, or Willow Tit?

5. Which of these is a UK 'Red' species due to the decline of its population – Brambling, Goldfinch, Linnet, or Siskin?

6. What caused the Dartford Warbler population to crash down to 10 pairs in the early 1960s?

7. What is Britain's commonest breeding bird?

8. What percentage of the worldwide Golden Eagle population does the UK have?

Bird Populations – Answers

1. Drought in their African wintering grounds.
2. Slavonian, by over 8 times more
3. 4000
4. Willow Tit
5. Linnet
6. Cold winters in the UK.
7. The Wren
8. About 1%

Birds in the Arts

1. Which Gull is heard on the opening music for the BBC Radio 4 show *Desert Island Discs*?

2. A discussion over which bird's speed led to the creation of the Guinness Book of Records?

3. Kehaar from *Watership Down* is what sort of bird?

4. Edgar Allen Poe wrote a poem about which bird?

5. H is for Hawk is about a lady and what bird?

6. The Mighty Sven in the film *Happy Feet 2* is what sort of bird?

7. Hedwig from the *Harry Potter* series is what sort of bird?

8. Vaughan Williams wrote about which bird ascending?

9. Which bird sang in Berkeley Square?

10. What was the surname of bird artist Charles Frederick?

Birds in the Arts – Answers

1. Herring Gull

2. Golden Plover

3. Black-headed Gull

4. The Raven

5. Goshawk

6. A Puffin

7. Snowy Owl

8. Skylark

9. A Nightingale

10. Tunnicliffe

Calls and Sounds

1. True or false – both male and female Bitterns boom?

2. Which wader makes a drumming sound through its tail feathers?

3. Which skulking wetland bird makes a rapid, loud, whip-like call?

4. This bird's summer call is often described as 'wet-me-lips' — which bird is it?

5. This bird is reputed to be saying 'little bit of bread and no cheese', but what is it?

6. Which two warblers have songs that have been compared to the sound of fishing reels?

7. Is a hooting Tawny Owl male, female, or could it be either?

8. Which type of bird might you confuse a Garden Warbler's song with – a Blackcap, Bullfinch, Greenfinch, or Wood Warbler?

Calls and Sounds – Answers

1. False – only males

2. Snipe

3. Spotted Crake

4. Quail

5. Yellowhammer

6. Grasshopper & Savi's

7. Male

8. Blackcap

Colour List

Make a list of all the colours that appear as a separate word at the start of a birds name (e.g. 'Red' for Red-necked Grebe, but nothing for Goldfinch, since it would need to be 'Gold Finch' to qualify). Words like 'Pied' are patterns rather than colours, so should not be included.

The answers have 20 colours listed.

Colour List – Answers

1. Black
2. Blue
3. Brown
4. Chestnut
5. Citrine
6. Cream
7. Green
8. Grey
9. Icterine
10. Indigo
11. Ivory
12. Magnolia
13. Olive
14. Purple
15. Red
16. Roseate
17. Rufous
18. Tawny
19. White
20. Yellow

1. Name 5 British birds whose name starts with 'Short'.

2. What is a Commic Tern?

3. Which bird is most common according to the RSPB's most recent Garden Birdwatch Survey?

4. Which bird has the longest legs, in relation to its body length?

5. What is the commonest letter of the alphabet for the start of English names on the British list? As an example, Quail would be Q.

6. Which two biologists have the most British birds named after them?

7. Name four British birds whose name start with the word 'Marsh'.

8. What is the common name for a female Swan?

General Knowledge – Answers

1. Short-billed Dowitcher, Short-eared Owl, Short-toed Eagle, Short-toed Lark, Short-toed Treecreeper.

2. Either an Artic or Common Tern.

3. House Sparrow (2017 survey)

4. Black-winged Stilt

5. S - 13% of names

6. Pallas (Pallas's Grasshopper Warbler, Pallas's Reed Bunting, Pallas's Sandgrouse, Pallas's Warbler) and Wilson (Wilson's Petrel, Wilson's Warbler, Wilson's Phalarope, Wilson's Snipe).

7. Marsh Harrier, Marsh Sandpiper, Marsh Tit, Marsh Warbler.

8. A pen.

Warbler Word Search

Q	B	S	R	D	V	W	Y	S	T
W	A	P	E	W	A	K	O	E	U
G	R	E	E	N	I	S	H	D	S
E	R	C	D	A	I	L	T	G	R
R	E	T	N	M	R	B	L	E	C
T	D	A	A	E	O	T	L	O	B
D	H	C	C	W	U	M	I	Q	W
Y	E	L	L	O	W	H	W	C	V
C	M	E	L	O	D	I	O	U	S
G	P	D	M	D	S	G	L	I	G

There are 10 Warbler names to find in the word search above. They can be found vertically, horizontally or diagonally, but not backwards.

Warbler Word Search – Answers

	B	S	R				S		
	A	P	E	W			E		
G	R	E	E	N	I	S	H	D	
	R	C	D	A		L		G	
	E	T			R		L	E	
	D	A				T		O	
		C		W			I		W
Y	E	L	L	O	W			C	
	M	E	L	O	D	I	O	U	S
		D		D					

1. Melodious Warbler
2. Greenish Warbler
3. Sedge Warbler
4. Spectacled Warbler
5. Reed Warbler
6. Wood Warbler
7. Yellow Warbler
8. Willow Warbler
9. Arctic Warbler
10. Barred Warbler

Black and White

What are these birds?

1. Small black and white bird. Summer Migrant. Likes Oak woodland in the west of Britain.

2. Small black and white bird with a red crown. Likes larger patches of undisturbed deciduous woodland.

3. Medium size black and white bird with red legs and long red beak. Likes estuaries in winter.

4. Medium size black bird with white patch on the wings and red feet. Seen on the sea off Scotland.

5. Pair of medium size black birds with white belly and wings. Long tail. Seen in parkland.

6. Black duck with small white patch by its eye. Sitting on the sea off of Scotland.

7. Black bird with white forehead. Diving in a reservoir.

8. Black bird with white breast band. Flying across moorland.

Black and White – Answers

1. Pied Flycatcher.

2. Lesser spotted Woodpecker.

3. Oystercatcher.

4. Black Guillemot.

5. Magpie.

6. Velvet Scoter.

7. Coot.

8. Ring Ouzel.

Anagrams

Work out the birds' names from the anagrams below.

1. Wool Sees Tree Fights Drone

2. Ram Stand In

3. Darn Leg Sin

4. Drowned So A Pip

5. If Elf Read

6. Odd Hero Cow

7. Lacquer Raw Bait

8. Henpecked Pearl Road

Anagrams – Answers

1. Lesser White-Fronted Goose
2. Sand Martin
3. Sanderling
4. Wood Sandpiper
5. Fieldfare
6. Hooded Crow
7. Aquatic Warbler
8. Red-necked Phalarope

Introductions

Which of these birds were introduced to Britain by people rather than occurring here naturally?

1. Little Owl

2. Egyptian Goose

3. Ring necked Parakeet

4. Collared Dove

5. Black Swan

6. Canada Goose

7. Golden Pheasant

8. Ruddy Duck

9. Eagle Owl

10. Mandarin Duck

11. Eagle Owl

12. Feral Pigeon

Introductions – Answers

All of them were human-introduced, apart from Collared Dove, which colonized naturally.

Jokes

1. What bird has criminal tendencies?

2. What bird is always out of breath?

3. Why does a Stork stand on one leg?

4. What do Geese eat?

5. Where do the toughest Magpies come from?

6. What does a curlew have in common with a meal at a fancy restaurant?

7. What birds do you find in a church?

8. What do pigeons drink?

9. What's the quietest bird?

10. What does an educated Tawny Owl say?

Jokes – Answers

1. A Robin.

2. A Puffin.

3. Because it would fall over if it lifted up the other one.

4. Gooseberries.

5. Hard boiled eggs.

6. Big bills

7. Birds of prey

8. Nest-café.

9. A stuffed bird.

10. Whom.

Famous Birdwatchers

Name the famous British bird watcher from the clues.

1. One of the Beatles.

2. Comedian and musician who appeared in *Black Books*.

3. Appeared in Gavin & Stacey.

4. Famous photographer from the 1960s onwards.

5. A Goodie

6. Famous for "playing all the right notes, but not necessarily in the right order" when conducted by Andre Previn.

7. Police Sergeant in the Bill.

8. Television petrol-head and writer.

9. Creator of the world's most famous spy.

10. This singer knows his arm from his Elbow.

Famous Birdwatchers – Answers

1. Paul McCartney
2. Bill Bailey
3. Alison Steadman
4. David Bailey
5. Bill Oddie
6. Eric Morecombe
7. Trudie Goodwin
8. Jeremy Clarkson
9. Ian Fleming
10. Guy Garvey

Abbreviations

What do these abbreviations and short hand terms refer to?

1. Ringtail

2. LBJ

3. Alcid

4. Bins

5. Corvid

6. Dipping

7. Blackwit

8. Acros

9. Pom

10. WeBS

Abbreviations – Answers

1. Female or immature of either Hen Harrier or Montagu's Harrier.
2. Little Brown Job – a small brown bird.
3. Birds from the Alcidae family e.g. Guillemot, Razorbill.
4. Binoculars.
5. A member of the crow family.
6. Missing out on a bird you had hoped to see.
7. Black-tailed Godwit.
8. Acrocephalus warblers like Reed and Sedge warbler.
9. Pomarine Skua.
10. Wetland Bird Survey.

Foreign Names

Match the English name to the name for the same bird used in other parts of the world.

1. Robin	Gerfalke (German)
2. Great northern Diver	Common loon (USA)
3. Golden Eagle	Cudyll Coch (Welsh)
4. Bittern	Scricciolo (Italian)
5. Black-throated Diver	Parelduiker (Dutch)
6. Nightjar	Rohrdrommel (German)
7. Oystercatcher	Iolaire-bhuidhe (Gaelic)
8. Wren	Rougegorge (French)
9. Gyrfalcon	Náttfari (Icelandic)
10. Kestrel	Ostraceiro (Portuguese)

Foreign Names – Answers

1. Robin   Rougegorge (French)

2. Great northern Diver   Common loon (USA)

3. Golden Eagle   Iolaire-bhuidhe (Gaelic)

4. Bittern   Rohrdrommel (German)

5. Black-throated Diver   Parelduiker (Dutch)

6. Nightjar   Náttfari (Icelandic)

7. Oystercatcher   Ostraceiro (Portuguese)

8. Wren   Scricciolo (Italian)

9. Gyrfalcon   Gerfalke (German)

10. Kestrel   Cudyll Coch (Welsh)

More True or False

1. Black guillemots are 'left-beaked' for holding fish.

2. A Puffin can hold up to 80 fish in his bill

3. Eider duck have different ways of opening mussels, which give them different beak shapes.

4. Coots are so aggressive they may kill their own chicks if they have too many to feed.

5. Slavonian Grebes often eat their own feathers.

6. Cormorants do not have waterproof plumage.

7. The Great Tit has the largest clutch of any bird species in Britain.

8. Canada Geese are reputedly amongst the most inedible of birds.

9. Egyptian Geese nest in winter.

10. The expansion of the Great Crested Grebe in Britain is probably due to the colonisation of Britain by Zebra Mussels.

More True or False – Answers

1. False - they are either left or right beaked.
2. True
3. False – it's true for Oystercatchers
4. True
5. True
6. True
7. False – It's the partridge
8. True
9. True
10. False – true for the Tufted Duck